Toys i

by Elizabeth Moore

Consultant:
Adria F. Klein, Ph.D.
California State University, San Bernardino

capstone
classroom

Heinemann Raintree • Red Brick Learning
division of Capstone

In the past, people made dolls from corn husks.

People made dolls from cloth too.

In the past, people made toy trucks from wood.

They made pull toys from wood.

They made spinning tops out of wood.

In the past, people made horses out of tin.

What are toys made of today?